THIRD
ASSESSMENT IN
MATHEMATICS

ANSWER BOOK

JM BOND

Nelson

Paper 1

Write the correct fraction in each space.

1 Shaded $\frac{5}{7}$

2 Unshaded $\frac{2}{7}$

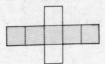

3 Shaded $\frac{1}{8}$

4 Unshaded $\frac{7}{8}$

5 Shaded $\frac{5}{12}$

6 Unshaded $\frac{7}{12}$

7 Shaded $\frac{3}{7}$

8 Unshaded $\frac{4}{7}$

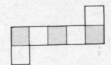

Put a sign in each space so that each sum will be correct.

9–10 $(3 \times 4) - 4 = 8$ 11–12 $(5 \times 4) + 4 = 24$

13–15 What numbers are coming out of this machine?

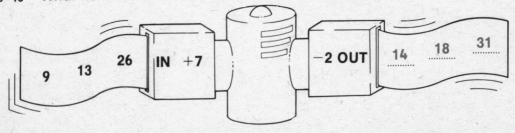

16–20 Complete these figures. The dotted line is the line of symmetry.

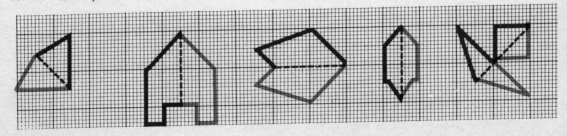

2

Here is a chart to show which colours we like.

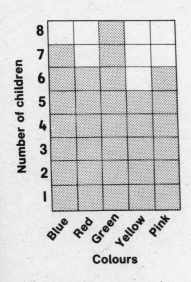

21 How many are there in our class? .... 32 ....

22 Which is the most popular colour? .... green ....

23 Which is the least popular colour? .... yellow ....

24–25 Which two colours are equally popular? .... red and pink ....

Half of those who like green are boys, and half of those who like red are girls. One boy likes pink, 6 boys like blue, and 3 girls like yellow.

26 How many boys are there in the class? .... 16 ....

27 How many girls? .... 16 ....

28 I think of a number, double it, add 5 and then divide it by 3. The answer is 7. What is the number? .... 8 ....

Write the next two numbers in each of the following rows:

29–30	1	2	4	8	16	32
31–32	800	400	200	100	50	25
33–34	9	12	15	18	21	24
35–36	55	50	45	40	35	30
37–38	48	42	36	30	24	18

3

Write as mixed numbers:

39 $\frac{7}{4} = 1\frac{3}{4}$ **40** $\frac{4}{3} = 1\frac{1}{3}$ **41** $\frac{5}{2} = 2\frac{1}{2}$ **42** $\frac{7}{3} = 2\frac{1}{3}$

43 $\frac{6}{5} = 1\frac{1}{5}$ **44** $\frac{9}{2} = 4\frac{1}{2}$

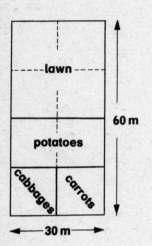

45 What fraction of the garden is lawn? $\frac{1}{2}$

46 What fraction is for cabbages? $\frac{1}{8}$

47 What fraction is for potatoes? $\frac{1}{4}$

48 What fraction is for carrots? $\frac{1}{8}$

49 What is the area of the garden?   1800 m²

50 What is the area of the lawn?   900 m²

Paper 2

1–6 Complete the following:

10 minutes × $\boxed{6}$ = 1 hour   20p × $\boxed{5}$ = £1.00

50p × $\boxed{4}$ = £2.00 $2\frac{1}{2}$ = $\boxed{5}$ halves

$1\frac{3}{4}$ = $\boxed{7}$ quarters   10p × $\boxed{50}$ = £5.00

7–12

−5

Input	8	14	32	40
Output	3	9	27	35

÷5

Input	25	35	55	60
Output	5	7	11	12

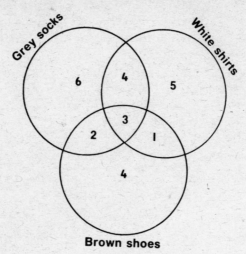

Grey socks   White shirts

6   4   5

3

2   1

4

Brown shoes

Some boys made this Venn diagram to show what they were wearing.

13   How many boys were there?   <u>25</u>

14–16   <u>15</u> wore grey socks, <u>13</u> wore white shirts and <u>10</u> wore brown shoes.

17   <u>5</u> boys wore both grey socks and brown shoes.

18   The number who wore both grey socks and white shirts was   <u>7</u>

19   How many boys wore both white shirts and brown shoes?   <u>4</u>

20   How many wore white shirts, grey socks and brown shoes?   <u>3</u>

21   The number of boys not wearing brown shoes was   <u>15</u>

22–23   <u>10</u> boys were not wearing grey socks, and <u>12</u> were not wearing white shirts.

Underline the correct answer to each sum.

24 $\frac{1}{2} + \frac{1}{4}$ =   <u>$\frac{3}{4}$</u> $\frac{1}{8}$ $\frac{3}{8}$ $\frac{1}{6}$

25 $1.00 - 0.9$ =   0.01   1.9   1.1   <u>0.1</u>

26 $2 \div \frac{1}{2}$ = $\frac{1}{4}$ 1   <u>4</u>   2

27 0.2×0.2 =   0.4   <u>0.04</u>   0.2   0.01

28 $\frac{1}{2} \times \frac{1}{3}$ = $\frac{2}{5}$ <u>$\frac{1}{6}$</u> $\frac{1}{5}$ $\frac{2}{6}$

29 What number when divided by 6 has an answer of
7 remainder 3?

45

30 Divide 25 metres into 8 equal pieces. How long is each
piece?

3.125 m

31–35 A regular shape has all its sides and angles equal. Underline the
shapes below which are regular.

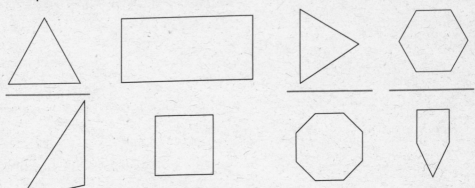

36–37 Which is the greater, and by how many
centimetres, 4.75 m or 389 cm?

4.75 m by 86 cm

38 Angle a = 55°

39 Angle b = 30°

40 Angle c = 60°

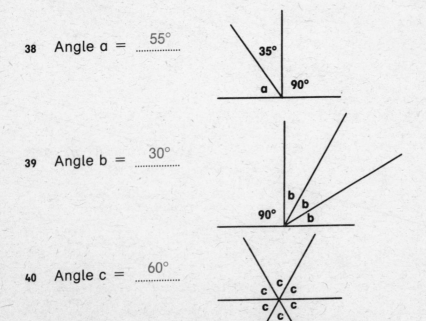

6

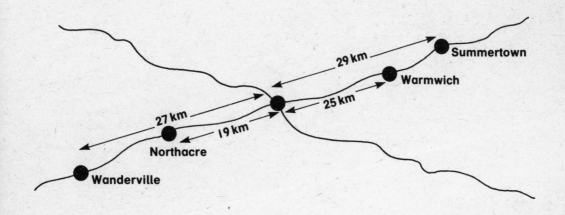

41	How far is it from Northacre to Summertown?	48 km
42	How far is it from Wanderville to Summertown?	56 km
43	How far is it from Warmwich to Northacre?	44 km
44	How far is it from Warmwich to Wanderville?	52 km

45–46 In a school of 300 children $\frac{3}{5}$ are boys. How many boys are there? 180 How many girls? 120

Write the following fractions as decimals:

47 $\frac{1}{10}$ 0.1   **48** $\frac{3}{100}$ 0.03   **49** $\frac{17}{100}$ 0.17   **50** $\frac{9}{10}$ 0.9

Paper 3

1 A train left Dover at 11.25 and arrived in London at 13.03. How long did the journey take?   1 hr 38 min

2 Here are 6 numbers. Find the largest and multiply it by the smallest.

32   27   35   33   26   29   910

3–5 Which of the amounts below could I make up with these five coins?

19p　　　28p　　　33p　　　29p　　　37p

Here is a pie-chart which shows how Dean spent last Tuesday.

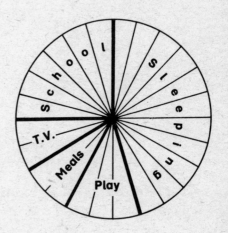

6–7 He was asleep for ___11___ hours and at school for ___6___ hours.

8–9 He watched T.V. for ___2___ hours and spent ___2___ hours eating.

10 Dean was playing for ___3___ hours.

11 What fraction of the day was Dean at school? ___$\frac{1}{4}$___

12 What fraction of the day did Dean spend playing? ___$\frac{1}{8}$___

13–15 What number is halfway between:

20 and 30　　　　21 and 35　　　　17 and 29

　　25　　　　　　　28　　　　　　　23

16 How many 250 g packets of biscuits can be made from 12 kg?　　　___48___

Multiply each of these numbers by 10.

17–21　3.4　　　14.6　　　0.78　　　2.68　　　5.9
　　　　　34　　　146　　　7.8　　　26.8　　　59

22–26　31.22　　0.56　　　2.5　　　0.27　　　28
　　　　　312.2　　5.6　　　25　　　2.7　　　280

8

You have a die (the sides are numbered 1 to 6)

27 How likely is it that you will throw an even number?   .....1 in 2.....

28 How likely is it that you will throw a 6?   .....1 in 6.....

29 If $\frac{3}{4}$ of my money is 27p, how much have I?   .....36p.....

30 If it takes Karen 17 minutes to walk to school, what time must she leave home to get to school by 8.50 a.m.?   .....8.33 a.m......

31 Paul and Timothy have each a piece of string. Paul's piece is 6.7 m long and Timothy's is 584 cm. Who has the longer piece?   .....Paul.....

32 How much longer is it?   .....86 cm.....

Here is a graph which shows the average number of people in Southwich who watch television every night.

———

shows those who watch BBC, and

----

shows those who watch ITV.

33–34 On which 2 nights do the same number of people watch BBC and ITV?   .....Thursday and Sunday.....

35 On which night is there the biggest difference in the number of people who watch BBC and ITV?   .....Friday.....

36 BBC has the fewest viewers on   .....Monday.....

37 ITV has the fewest viewers on   .....Wednesday.....

38 How many more people watch ITV than BBC on Saturdays?   .....500.....

39 I gave Andrew $\frac{1}{2}$ of my money. A half of what I had left was 15p. How much had I at first?   .....60p.....

40	Line A measures	4 cm	A _____
41	Line B measures	5.5 cm	B _____
42	Line C measures	4.5 cm	C _____
43	Line D measures	2 cm	D _____

If, on a map, 1 cm represents 10 km:

44   Line A would represent   40 km

45   Line B would represent   55 km

46   Line C would represent   45 km

47–48   Fill in the missing figures in the following sums.

$$5 \text{ rem } 3$$
$$5 \overline{)\ 28}$$

$$6 \text{ rem } 2$$
$$6 \overline{)\ 38}$$

49   If my car uses 1 litre of petrol every 10 km, how many litres will I need to go 105 km?   10.5 litres

50   $\frac{3}{5}$ of a class of 25 are boys. How many girls are there?   10

Paper 4

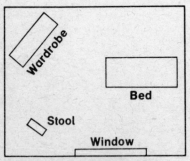

Scale : 1 cm represents 1 m

1–2   My bedroom is   5 m   long and   4 m   wide.

3–4   The window is   2 m   long and the bed is   2 m   long.

5–6   The wardrobe is   1.5 m   long and the stool is   50 cm   long.

7 How many times can I take 8 from 200?   25

8 Take the product of 6 and 7 from 50.   8

9 At Markswell school there are 13 classes and
   286 children. If the classes are all the same size, how
   many children are there in each class?   22

10–14 Match these digital and ordinary clocks.

15–20 Write one of these signs < (less than), > (more than),
or = in the spaces below.

$6 \times 6 \; > \; 17 + 18$ $3^2 \; < \; 2 \times 5$

$8 + 10 - 3 \; = \; 5 \times 3$ $\frac{3}{4}$ hour $>$ 40 seconds

$0.5 \text{ m} \; = \; 50 \text{ cms}$ $7 + 13 \; < \; 6 + 7 + 8$

21 Share £37.50 amongst 25 boys. They would have   __£1.50__   each.

22 What fraction of the square
    is covered with dots?   ..... $\frac{1}{9}$ .....

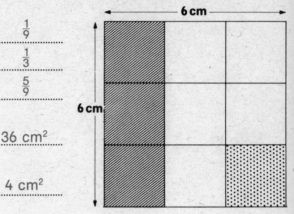

6 cm

6 cm

23 What fraction is shaded?   ..... $\frac{1}{3}$ .....

24 What fraction is white?   ..... $\frac{5}{9}$ .....

25 What is the area of the
    whole figure?   __36 cm²__

26 What is the area of the
    dotted square?   __4 cm²__

27 What is the area of the
    shaded squares?   __12 cm²__

28 What is the area of the white squares?   __20 cm²__

At school we grew some beans. John's was 0.70 m tall, Lisa's was $\frac{3}{4}$ metre tall and Kate's was 65 cm tall.

29 Who grew the tallest bean?   __Lisa__

30 Who grew the 2nd tallest?   __John__

31 What was the difference in height between Kate's bean
    and John's bean?   __5 cm__

Use these 2 sums to help you answer the following questions:

32 – 36

```
  357        712
+ 345      − 368
─────      ─────
  702        344
```

$368 + 344 =$ __712__

$702 − 345 =$ __357__

$712 − 368 =$ __344__

$702 − 357 =$ __345__

$712 − 344 =$ __368__

37 – 40 Here is part of a table chart. Can you complete it?

5 6	4 2	4 8	8 8
6 4	4 8	6 0	9 9
7 2	5 4	7 2	1 1 0

41 $\frac{2}{10} + \frac{7}{10} - \frac{1}{10}$

$\frac{8}{10}$

42 $\frac{9}{10} - \frac{6}{10} + \frac{2}{10}$

$\frac{5}{10}$

43 $\frac{9}{10} + \frac{9}{100}$

$\frac{99}{100}$

44 $\frac{24}{100} + \frac{13}{100} - \frac{8}{100}$

$\frac{29}{100}$

45 $\frac{29}{100} - \frac{16}{100} + \frac{14}{100}$

$\frac{27}{100}$

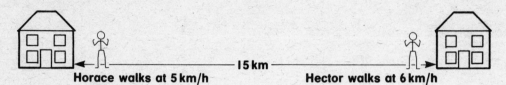

Horace walks at 5 km/h — 15 km — Hector walks at 6 km/h

46 How long would it take Horace to walk to Hector's house?

3 hours

47 How long would it take Hector to walk to Horace's house?

$2\frac{1}{2}$ hours

48 If Hector left his house at 5.45 p.m. when would he get to Horace's house?

8.15 p.m.

49 If the sum of two numbers is 26, and their difference is 4, what is the smaller number?

11

50 What is the larger number?

15

Paper 5

1–7 What is the first number after 123 to end in

one zero 130   2 zeros 200   3 zeros 1000

Start at 30 and count on in 35s. Don't go beyond 200.

30   65   100   135   170

8–12 Match the following:

23.10   D   A   5 minutes before noon

16.05   E   B   Wake up, have breakfast

11.55   A   C   School starts

08.50   C   D   Nearly midnight

07.30   B   E   Tea-time – home from school

Write the next two numbers in each of the following lines:

13–14	15	20	30	35	45	50 60
15–16	$3\frac{1}{2}$	5	$6\frac{1}{2}$	8		$9\frac{1}{2}$ 11
17–18	1.85	1.90	1.95			2.00 2.05
19–20	2	5	7	10	12	15 17
21–22	50	40	39	29	28	18 17
23–24	1234	123.4	12.34	1.234		0.1234 0.01234

25 Divide 4.675 by 5 _0.935_

26–31 Round off the following numbers to the nearest 10.

17	39	52	75	66	28
20	40	50	80	70	30

32–37 Make out a timetable for Class 3B. They have five 30-minute lessons each morning. They have 15 minutes playtime after the 3rd lesson.

	Begins	Ends
1st lesson	09.05	9.35
2nd lesson	09.35	10.05
3rd lesson	10.05	10.35
Playtime	10.35	10.50
4th lesson	10.50	11.20
5th lesson	11.20	11.50

38 How long does this class spend at lessons each morning?   2 hr 30 min

39 Andrew has 54p, and Oliver has half as much as Andrew. How much have they together?   81p

40 If the distance all round a square is 22 cm, what is the length of each side?   5.5 cm

Look at the letters below. They form a pattern.

41 a b c d e a b c d e . . .   What would the 18th letter be?   c

42 x y z x y z . . .   What would the 10th letter be?   x

43 How many times can I take 28 from 756?   27 times

44 478 + 199 = 677   If I took 100 away from each of the 2
                             numbers the answer would be:

                             The same   100 less   200 less

45 478 − 199 = 279   If I took 100 away from each of the 2
                             numbers the answer would be:

                             The same   100 less   200 less

46 Add together $\frac{7}{8}$ and $\frac{1}{2}$ $1\frac{3}{8}$

47 How many quarters are there in $7\frac{3}{4}$?   31

48 In a school $\frac{3}{5}$ of the children were boys, and there were
    240 girls. How many boys were there?   360

49 How many children were there in the school?   600

50 How many days are there in the first 3 months of a Leap
    Year?   91

Paper 6

Write the following decimals as fractions in their lowest terms:

1 0.1 $\frac{1}{10}$ 2 0.03 $\frac{3}{100}$ 3 3.5 $3\frac{1}{2}$

4 2.2 $2\frac{1}{5}$ 5 5.25 $5\frac{1}{4}$ 6 9.75 $9\frac{3}{4}$

7–10 Underline the sums below which will have a remainder of 2.

        64 ÷ 8   44 ÷ 6   100 ÷ 5   82 ÷ 10

        38 ÷ 4   15 ÷ 5   74 ÷ 12   27 ÷ 4

11 What size is the smaller angle between 12 and 3? ...90°...

12 What size is the smaller angle between 2 and 3? ...30°...

13 The smaller angle between 4 and 6 is: ...60°...

14 The smaller angle between 6 and 9 is: ...90°...

15 The angle between 6 and 12 is: ...180°...

16 The smaller angle between 7 and 8 is: ...30°...

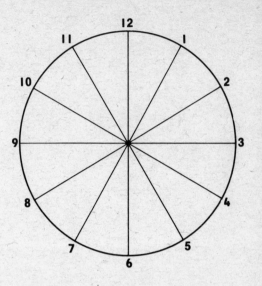

17 – 20

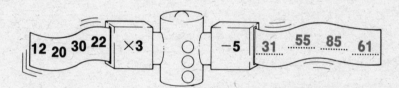

	Train A	Train B	Train C (Express)
Anyville	09.30	10.50	11.30
Bobtown	09.55	11.15	
Citywich	10.05	11.25	
Downderryo	10.15	11.35	
Earlham Road	10.40	12.00	
Fardon	11.15	12.35	13.00

21 – 26 Complete the timetable. Train B stops at the same stations as Train A and travels at the same speed. Train C takes 15 minutes less to complete the journey.

27 How long does the complete journey of Train A or Train B take? ...1 hr 45 min...

16

28   If a coach travels 315 km in 7 hours what is its average speed?   <u>45 km/h</u>

29   Take 47 from 102, and then divide your answer by 5.   <u>11</u>

Brand Y custard powder is obtainable in 3 sizes. A 500 g tin costs 85p, a 250 g tin costs 44p and a 125 g tin costs 24p.

30   What would you save if you bought one 250 g tin instead of two 125 g tins?   <u>4p</u>

31   How much cheaper is it to buy a 500 g tin than four 125 g tins?   <u>11p</u>

32   The difference between two numbers is 8. The smaller number is 17. What is the other?   <u>25</u>

33   The London train, due at 11.57 a.m., is 14 minutes late. At what time will it arrive?   <u>12.11 p.m.</u>

34   We had to guess the number of peas in a bottle. There were 700. Steven said there were 681, Robin said 715 and Jake 693. Who was the nearest?   <u>Jake</u>

35   Who was the next nearest?   <u>Robin</u>

36–38   Fill in the missing figures.

$$
\begin{array}{r} 8 \text{ Rem } 2 \\ 6\overline{)\underline{50}} \end{array}
\qquad
\begin{array}{r} 9 \text{ Rem } \underline{5} \\ 7\overline{)68} \end{array}
$$

39   School starts at 8.55 a.m. One day Katie was a quarter of an hour late. When did she arrive?   <u>9.10 a.m.</u>

40   There are 720 books in the school library. How many books has John read if he has read one-twelfth of them?   <u>60</u>

41   The sum of the ages of Wayne and Annette is 20. If Wayne is 4 years older than Annette, how old is he?   <u>12 years</u>

42   How old is Annette?   <u>8 years</u>

43 Multiply 36 by 200 ....7200.........

Put signs in the following sums to make them correct.

44–45 (5 ..+.. 9) ..−.. 8 = 6   **46–47** (12 ..÷.. 4) ..+.. 3 = 6

48–49 (6 ..×.. 2) ..+.. 4 = 16

50 What number is 12 more than 19? ....31.........

Paper 7

Some children were asked which instruments they played, so they made this Venn diagram.

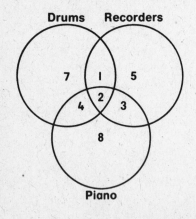

Drums   **Recorders**

7   1   5

4   2   3

8

Piano

1 How many children were asked? ....30.........

2 How many play the drums? ....14.........

3 How many play recorders? ....11.........

4 ...17... play the piano.

5 How many play both drums and recorders? ....3.........

6 How many play the piano and the drums? ....6.........

7 ...5... children play the recorder and the piano.

8 How many can play all three instruments? ....2.........

9 ...16... children don't play the drums.

10 Add half of 8 to twice 6. ....16.........

18

11 What is the area of
this square? _16 cm²_

12 What is the area of
the grey part? _4 cm²_

13 What is the total area of
the white parts? _8 cm²_

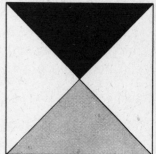

At Ladybird School there are 30 children in each of the 4 classes. Here is a pictogram which shows the children who can swim.

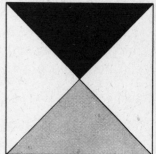 = 10 children who can swim		Number who can swim	Number who cannot swim
Class 1	🐟🐟	20	10
Class 2	🐟🐟🐟	25	5
Class 3	🐟	10	20
Class 4	🐟🐟	15	15

14–21 Fill in the spaces.

22 What fraction of the whole school can swim? _$\frac{7}{12}$_

23 What fraction of the whole school cannot swim? _$\frac{5}{12}$_

24 In which class can 50% of the children swim? _Class 4_

25 In which class can $\frac{2}{3}$ of the children swim? _Class 1_

26 In which class do the children who cannot swim
outnumber those who can by 2:1? _Class 3_

27–29 How many halves are there in:

$7\frac{1}{2}$ $10\frac{1}{2}$ 23

_15_   _21_   _46_

30 How many times can I subtract 0.05 from 5? _____100_____

There are 96 roses in a bed in the park. Half of them are yellow, $\frac{1}{4}$ are red, and the rest are pink.

31–33 __24__ roses are red, __48__ are yellow and __24__ are pink.

34–36 $3 \times \triangle = 12$ $\bigcirc + 2 = 5$ $4 \times \triangledown = 16$

so $\triangle = $ __4__ so $\bigcirc = $ __3__ so $\triangledown = $ __4__

37 In our library there are 200 books. $\frac{2}{5}$ of them are non-fiction. How many non-fiction books are there? _____80_____

38 How many fiction books are there? _____120_____

39 $\frac{3}{5}$ of the books are fiction. (Answer as a fraction.)

40 If I walk at a steady rate of 6 km per hour, how long will it take me to walk 15 km? $2\frac{1}{2}$ hours

41 If I keep up this rate, how far will I walk in $3\frac{1}{2}$ hours? _____21 km_____

42 There are 26 children in Class 1, and 30 children in Class 2. If 4 children leave Class 2, what will be the average number in both classes? _____26_____

43 My watch loses 2 minutes in 24 hours. If it is put right at 10 a.m. on Monday, what time will show on it at 10 p.m. on Monday? _____9.59_____

44 What time will show on it at 10 a.m. on Tuesday? _____9.58_____

45 What time will show on it at 10 a.m. on Wednesday? _____9.56_____

46 What time will show on it the following Monday at 10 a.m.? _____9.46_____

47–48 Joe and Sharon shared 24 books. Joe had 2 more than Sharon, so he had __13__ and Sharon had __11__

49 How many minutes are there between noon and 13.29? _89_

50 How many weeks are there in 266 days? _38_

Paper 8

Write the correct fraction in each space.

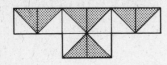
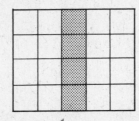
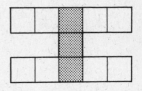

1-2 Shaded $\frac{1}{2}$ **3-4** Shaded $\frac{1}{5}$ **5-6** Shaded $\frac{3}{11}$

Unshaded $\frac{1}{2}$ Unshaded $\frac{4}{5}$ Unshaded $\frac{8}{11}$

Find the order of rotational symmetry for each of these shapes.

7-10

_4_ _2_ _1_ _3_

11-17 Here are the temperatures in some cities one day last winter.

Find the difference in temperature between:	

Barcelona	14
Bombay	31
Dublin	7
Helsinki	-3
Zurich	-2
Moscow	-11
Oslo	0

Find the difference in temperature between:

Bombay and Dublin	24
Zurich and Barcelona	16
Moscow and Zurich	9
Oslo and Bombay	31
Helsinki and Oslo	3
Dublin and Zurich	9
Barcelona and Moscow	25

£	kg	l
7.49	1.789	1.682
+ 8.56	+ 2.567	+ 2.379
16.05	4.356	4.061

21 Tony cycles at an average speed of 18 km/h. If he can cycle to school in 10 minutes, how far from school does he live?

_____3 km_____

22 One day he walked to school at 6 km/h. It took him

$\frac{1}{2}$ hour

23–27 Put a decimal point in each of the following numbers so that the 3 has a value of $\frac{3}{10}$.

1 2.3 4   1 4 2.3   1.3 2 4   .3 2 4 1   4.3 2 1

28–33 Write one of the signs $<$ $>$ $=$ in each space.

$4 + 7 - 1 = 2 \times 5$ $5 \times 9 > 9 + 5$ $30 \div 6 < 3 \times 2$

$\frac{1}{2}$ hour $> 4 \times 7$ mins $3 l < 4000$ ml 150 cm < 2 m

34 What is the perimeter of a room which is 3.2 m long and 2.5 m wide?

_____11.4 m_____

Here are the birth dates of some children.

A   30.12.88   B   7.8.87   C   29.10.87

D   24.12.87   E   20.5.88   F   29.2.88

35 Who is the oldest?

B

36 Who is the youngest?

A

37 Who has a birthday during the summer holidays?

B

38 Who has a birthday only once every four years?

F

39 Whose birthday is in October?

C

40 Who has a birthday on Christmas Eve?

D

41 If Emma was 3 cm taller she would be twice the height of her brother. He is 75 cm tall. How tall is Emma?

_____1.47 m_____

22

This pie chart shows the favourite sports of 32 children.

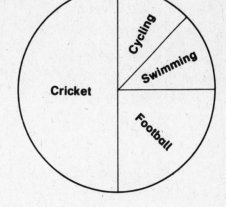

42 How many children prefer football? ....8.......

43 How many children prefer swimming? ....4.......

44 How many children prefer cricket? ....16.......

45 How many children prefer cycling? ....4.......

46 What fraction of the class prefers football? ....$\frac{1}{4}$.......

47–48 Which two sports are equally popular? ....swimming and cycling.......

49 What fraction of the class likes cricket best? ....$\frac{1}{2}$.......

50 What fraction of the class likes cycling best? ....$\frac{1}{8}$.......

Paper 9

Look at the pattern below.

1–3 1a2b3c1a2b3c . . . What would the 20th number or letter be? ....a.......

9g8f7e6d9g . . . What would the 23rd number or letter be? ....6.......

What would the 40th be? ....d.......

There are 240 books in our library. $\frac{1}{10}$ of them are about sport, $\frac{1}{8}$ are about nature, $\frac{1}{6}$ are travel stories, $\frac{1}{5}$ are adventure stories, and $\frac{1}{3}$ are school stories.

4–5 ....40....... books are about travel and ....80....... are school stories.

6–7 ....30....... are about nature and ....48....... are adventure stories.

8 How many books about sport are there? ....24.......

9 How many other books are in the library? ....18.......

23

10 Take forty-nine from two thousand and ten. _____ 1961

Here are some thermometers which show the temperatures at midday each day for one week last winter.

11–13 On which days was the temperature lower than the day before?

Tuesday, Wednesday and Friday

14–16 On which days was the temperature higher than on the previous day?

Thursday, Saturday and Sunday

17 Which day had the lowest temperature? _____ Friday

18 Which day had the highest temperature? _____ Monday

19–20 Between which consecutive days was the biggest difference in temperature? _____ Thursday and Friday

21–22 Between which consecutive days did the temperature change least? _____ Saturday and Sunday

23 Find the average midday temperature for the week. _____ 5°C

24 How many days had temperatures above this average? _____ 3

25 How many square metres of carpet would be needed to cover a floor 5 m long and 4.5 m wide? _____ 22.5 m²

26 What would this cost if the carpet was £8.00 a square metre? _____ £180.00

27 How many tiles, each 20 cm × 30 cm, would be needed to cover a floor 5 m × 6 m? _____ 500

28 How many days are there in spring (March, April and May)? _____ 92

29 Find the length of a queue of 21 cars (average length 4 m). There is a space of 1 metre between each pair of cars. _____ 104 m

30 Into how many squares the size of square A could the large figure be divided?

9

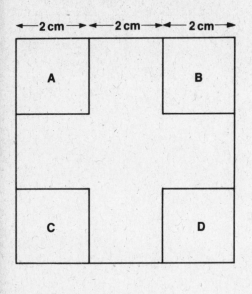

←—2 cm—→ ←—2 cm—→ ←—2 cm—→

A B

C D

Write the area of:

31 Square A .... 4 cm²

32 Squares B and C .... 8 cm²

33 Square D .... 4 cm²

34 The large square .... 36 cm²

35 The cross .... 20 cm²

36 What is the perimeter of the large square? .... 24 cm

37 What is the perimeter of square C? .... 8 cm

38–40 There are 10 balls in a bag. They are numbered 1 to 10.

What is the chance of picking out an even numbered ball? .... 1 in 2

a number which can be divided by 5? .... 2 in 10 or 1 in 5

a number which can be divided by 3? .... 3 in 10

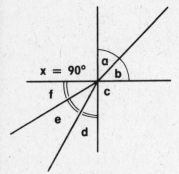

x = 90°

Angle a = Angle b
Angle f = Angle e = Angle d

41 Angle a = ... 45°   **42** Angle b = ... 45°

43 Angle c = ... 90°   **44** Angle d = ... 30°

45 Angle e = ... 30°   **46** Angle f = ... 30°

47 Angle a + b + c + d + e + f + x = ... 360°

48 The product of 2 numbers is 42. One of the numbers is 7. What is the other number? 6

49 The distance all round a room is 24 metres. The room is twice as long as it is wide. Its length is: .... 8 m

50 Its width is: .... 4 m

Paper 10

Write the correct fraction in each space.

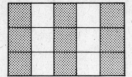  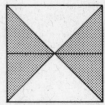

1–2 Shaded $\frac{3}{5}$ **3–4** Shaded $\frac{4}{5}$ **5–6** Shaded $\frac{1}{2}$

Unshaded $\frac{2}{5}$ Unshaded $\frac{1}{5}$ Unshaded $\frac{1}{2}$

Put a sign in each space so that each sum will be correct.

7–8 (5 $+$ 2) $-$ 4 = 3 **9–10** (8 $-$ 1) $+$ 2 = 9

11–12 (8 $-$ 3) $\times$ 2 = 10 **13–14** (3 $+$ 6) $-$ 4 = 5

15–19 Arrange the following numbers in order, starting with the smallest.

2.111   2.1   2.01   2.11   2.101

2.01   2.1   2.101   2.11   2.111

20 When a bottle is $\frac{3}{4}$ full it holds 6 litres. How much does it hold when it is $\frac{1}{2}$ full?

4 litres

21 Multiply 456 by 30

13 680

22 How many glasses holding $\frac{1}{4}$ litre could be filled from a bottle holding 2.5 litres?

10

23 What is the smallest number which must be added to 368 to make it exactly divisible by 27?

10

24 How many squares 5 cm × 5 cm can I cut from a card 20 cm × 40 cm?

32

25 How much have I left from £3.00 after buying 4 bunches of flowers at 65p each?

40p

26 I went to the cinema at 2.50 p.m. and came out at 5.30 p.m. How long was I there?

2 hr 40 min

Name	English	Maths	History	Total marks
Jackie	47	65	48	160
Gary	80	77	78	235
Liz	73	70	74	217
Jane	90	57	64	211

27 Who had the highest marks in any exam? — Jane

28 Who had the lowest marks in any exam? — Jackie

29 Who had the highest total marks? — Gary

30 Who had the 2nd highest total marks? — Liz

31 Who had the 3rd highest total marks? — Jane

32 Who had the lowest total marks? — Jackie

33 Who gained over 75 marks in three exams? — Gary

34 Who gained fewer than 50 marks in two exams? — Jackie

35 Who was top in English? — Jane

36 Who was 3rd in maths? — Jackie

37 Who was 2nd in maths? — Liz

38 I go to bed at 7.30 p.m., and get up at 7.15 a.m. How long am I in bed? — 11 hr 45 min

39 There are 28 in our class. One day there were 6 times as many present as there were absent. How many were present? — 24

40 How many were absent? — 4

41–48 Write in quarters:

$2\frac{1}{4} = \frac{9}{4}$ $1\frac{3}{4} = \frac{7}{4}$ $2\frac{3}{4} = \frac{11}{4}$ $1\frac{1}{4} = \frac{5}{4}$

$3 = \frac{12}{4}$ $4\frac{1}{4} = \frac{17}{4}$ $3\frac{3}{4} = \frac{15}{4}$ $4\frac{3}{4} = \frac{19}{4}$

49 Kim has 75p and Michael has 91p. How much must Michael give Kim so that they will both have the same amount? — 8p

50 How much will they each have then? — 83p

Paper 11

1   Travelling at an average speed of 60 km/h we went 150 km.
    If we left home at 11.30 at what time did we arrive?   ......14.00......

2–5   Shop A had 1 litre of bleach for £1.80, B had 250 ml for 55p,
      Shop C had $\frac{3}{4}$ litre for £1.47 and Shop D had 500 ml for £1.05.

      Shop ...A... was the cheapest   Shop ...C... was the 2nd cheapest
      Shop ...D... was the 3rd cheapest   Shop ...B... was the dearest

6–13
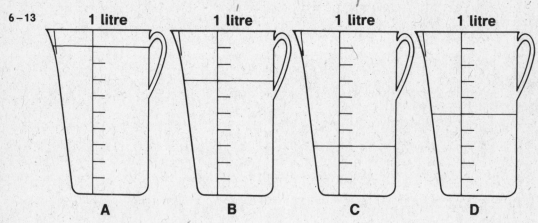

A contains ...900 ml...

B contains ...700 ml...

C contains ...300 ml...

D contains ...500 ml...

How much would I have to add to each jug to make it hold 1 litre?

100 ml   300 ml   700 ml   500 ml

14–17   Fill in the missing numbers.

          7 rem 3   5 rem 4   5 rem 3   4 rem 3
       8) 5 9   9) 4 9   7) 3 8   9) 3 9

18–21   Write down all the numbers between 30 and 60
        which are exactly divisible by 8.   32, 40, 48, 56

22-24 Write down all the numbers between 30 and 50 which are exactly divisible by 7.
_35, 42, 49_

25-28 Put a decimal point in each of the following numbers so that the 7 will have the value of 7 units.

8 9 7.5 6   5 7.9 5 6   7.9 8 5 6   9 8 6 7 5

Write the next two numbers in each of the following lines.

29-30	81	72	63	54	_45_ _36_
31-32	8	4	2	1	_$\frac{1}{2}$_ _$\frac{1}{4}$_
33-34	1.25	1.50	1.75		_2.00_ _2.25_
35-36	$7\frac{1}{4}$	$8\frac{1}{2}$	$9\frac{3}{4}$		_11_ _$12\frac{1}{4}$_
37-38	25	20	16	13	_11_ _10_
39-40	20	21	23	26	_30_ _35_

41 Mary is knitting some teddy bears for the school fete. She will need 5 balls of wool which cost £1.55 each. This will cost
_£7.75_

42 She will also need some knitting needles, also at £1.55, and a pattern costing 95p. The total cost will be
_£10.25_

43-46

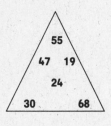

Which of the numbers in the triangle:

Has a remainder of 4 when divided by 8? _68_

Can be divided exactly by 5 and 6? _30_

Has a remainder of 3 when divided by 8? _19_

Has a remainder of 2 when divided by 5? _47_

There are 24 in our class. $\frac{3}{8}$ of the children are girls.

47-48 There are _9_ girls and _15_ boys.

49 How many times does 20 divide into 340? _17_

50 There were 36 black cats and 12 tabby cats. What fraction of the cats were tabby?
$\frac{1}{4}$

1 How many packets, each containing 50 g, could be filled
   from $3\frac{1}{2}$ kg?   70

2 A train departs at 07.49 and arrives at 10.13.
   How long is the journey?   2 hr 24 min

A mixed number is a whole number together with a fraction, such as $1\frac{1}{2}$.

3–5 Turn these into mixed numbers.

   $\frac{13}{10}$ $1\frac{3}{10}$ $\frac{7}{5}$ $1\frac{2}{5}$ $\frac{8}{3}$ $2\frac{2}{3}$

6–8 Multiply these numbers by 10.

   0.024   0.0076   0.75
   0.24   0.076   7.5

Here is a plan of our bathroom.
2 cm represents 1 m

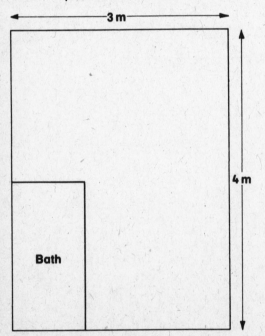

9 What is the
   perimeter of the
   bathroom?   14 m

10 We would need 60 floor tiles
    of 40 cm × 50 cm.

11 We decide not to
    put tiles under the
    bath. How many
    do we need now?   50 tiles

12 What is the area
    of the bath?   2 m²

13 What is the area
    of the bathroom?   12 m²

There are 30 in Class 3. Four out of ten of the children are girls.

14–15 There are 12 girls and 18 boys.

16 I get home from school at 4.25 p.m. Joanna gets home 45 minutes later than I do. When does Joanna get home? _5.10 p.m._

17 If Samantha saves 75p each week, how long will it take her to get enough money to buy a camera costing £18.00? _24 weeks_

18 Tom spent $\frac{5}{6}$ of his money. If he had 14p left, how much had he at first? _84p_

19
$$\begin{array}{r} 0.3 \\ \times 0.2 \\ \hline 0.06 \\ \hline \end{array}$$

20
$$\begin{array}{r} 0.8 \\ \times 0.7 \\ \hline 0.56 \\ \hline \end{array}$$

21
$$\begin{array}{r} 0.7 \\ \times 0.5 \\ \hline 0.35 \\ \hline \end{array}$$

The children in our class made this Venn diagram to show how many of them have fair hair and blue eyes.

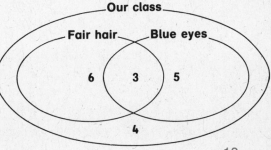

22 How many children are in our class? _18_

23 How many children have fair hair? _9_

24 How many have blue eyes? _8_

25 How many have both blue eyes and fair hair? _3_

26 _9_ children don't have fair hair.

27 _10_ children don't have blue eyes.

Fill in the missing figures:

28 $5 \times 6 \times \underline{10} = 300$

29 $3 \times 4 \times \underline{20} = 240$

30 $7 \times 2 \times \underline{4} = 56$

31 $8 \times 3 \times \underline{4} = 96$

32 $5 \times 8 \times \underline{5} = 200$

33 $10 \times 3 \times \underline{9} = 270$

34 3×0.007
_0.021_

35 4×1.01
_4.04_

31

Turn these fractions into percentages.

36 $\frac{20}{40} = $ 50% **37** $\frac{5}{25} = $ 20% **38** $\frac{40}{40} = $ 100% **39** $\frac{20}{25} = $ 80%

40 $\frac{10}{40} = $ 25% **41** $\frac{15}{25} = $ 60%

42 There were 42 chocolates in a box. Mary and her two brothers shared them equally, and made them last a week. How many chocolates did they each have on one day?

2

43 A train leaves Yarborough at 0930 and travels to Narborough, 480 km away. If the average speed of the train is 80 km/h, what time should it arrive in Narborough?

1530

44 At a sale the goods were reduced by 12p in the £1.00. How much did I pay for an article which cost £4.00 before the sale?

£3.52

We have lessons from 9 to 12 o'clock in the morning, and from 2 to 4 p.m. We have two breaks each day of 15 minutes each.

45-46 We work $4\frac{1}{2}$ hours a day, and $22\frac{1}{2}$ hours a week.

47-50

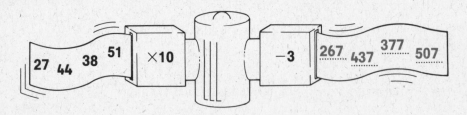

27 44 38 51 ×10 −3 267 437 377 507

Paper 13

1-4 Make these figures symmetrical. The dotted line is the line of symmetry.

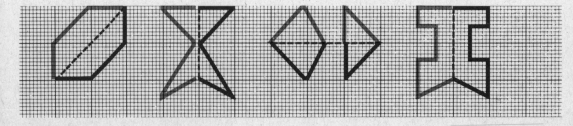

5 The sum of two numbers is 31. If one number is 15, what is the other?   16

The children in Class 3 drew this chart to show where they live.

High Street	👤	👤	👤	👤	👤	👤	👤	👤	👤	
Shore Drive	👤	👤								
Bay Avenue	👤	👤	👤	👤						
Low Road	👤	👤	👤	👤	👤	👤	👤			
Abbey Crescent	👤	👤	👤	👤	👤					

6 How many children live in High Street?   9

7 How many children live in Abbey Crescent?   5

8 How many children live in Shore Drive?   2

9 How many more children live in Low Road than in Bay Avenue?   3

10 What fraction of the class lives in High Street? $\frac{1}{3}$

	Begins	Ends
Maths	09.15	10.00
English	10.00	10.40
Break	10.40	10.55
Games	10.55	11.30
Science	11.30	12.10

11 The maths lesson is 45 minutes long.

12 The science lesson lasts 40 minutes.

13 We do games for 35 minutes.

14 From 09.15 to 12.10 is 2 hours 55 minutes.

15 Do we have more time for lessons before break or after break?

<u>before</u>   after

16 How much longer?   10 minutes

```
    98        So 17    198        18    398        19    498
  + 76          + 176          +  76          + 176
  ----          -----          -----          -----
   174           374            474            674
```

33

20–23 Underline the shapes which have been divided symmetrically by the dotted line.

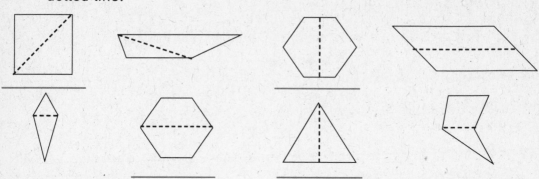

Fill in the missing figures.

24 5 minutes × $\boxed{12}$ = 1 hour **25** 20 cm × $\boxed{5}$ = 1 metre

26 125 ml × $\boxed{8}$ = 1 litre **27** 50 m × $\boxed{20}$ = 1 km

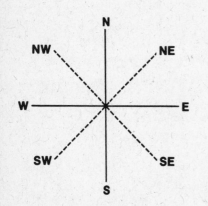

28 What is the size of the smaller angle between N and E? .... 90° ....

29 How many degrees are there between N and S? .... 180° ....

30 How many degrees are there between E and W? .... 180° ....

31 How large is the smaller angle between N and W? .... 90° ....

32 What is the size of the smaller angle between W and SW? .... 45° ....

33 What is the size of the smaller angle between N and NE? .... 45° ....

34 The smaller angle between N and SE = ... 135° ....

35 What is the size of the smaller angle between S and NW? .... 135° ....

36 My watch loses a minute each day. If I put it right at noon on Monday, what time will show on my watch at noon on Thursday? .... 11.57 ....

37 Multiply 20 by itself. .... 400 ....

38 How much larger is 740 than 74? .... 666 ....

39 How many weeks are there in 161 days? .... 23 ....

40 How long is the church? ..90 m..

41–42 The churchyard is ..150 m.. long and ..90 m.. wide.

43 School Lane is ..240 m.. long.

44–45 The school is ..120 m.. long and ..60 m.. wide.

46–47 The playground is ..150 m.. long and ..90 m.. wide.

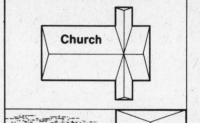
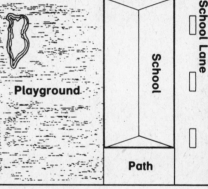

Scale: I cm represents 30 m

Church

School Lane

School

Playground.

Path

48 The bus was due at 11.57 a.m. It was 9 minutes late so it arrived at: ..12.06 p.m...

49 Tom delivers newspapers each weekday (not Sunday). If he delivers 49 each day, how many does he deliver in a week? .... 294 ....

50 If it takes him 40 minutes each day, how long does he spend delivering papers each week? ....4 hours....

Paper 14

1–4 Make these numbers 100 times as large:

4.8	0.24	37.5	27
480	24	3750	2700

	Train A	Train B	Train C (Express)
Midwich	11.17	13.05	14.20
Moreton	11.22	13.10	
Marston	11.32	13.20	
Middlewich	11.47	13.35	
Marchester	12.07	13.55	
Muddlecombe	12.22	14.10	15.20

5–10 Complete the timetable. Train B stops at the same stations as Train A and travels at the same speed. Train C takes 5 minutes less to complete the journey.

11 How long does the complete journey of Train A or Train B take?

1 hr 5 min

12 I have an expanding curtain rod, 76 cm long. It stretches to $1\frac{1}{2}$ times its length. How long is it when stretched?

_1.14 m_

13 A piece of rope, 2.43 m, is cut into 9 equal pieces. How long is each piece?

_27 cm_

14 $3 - 1\frac{9}{10} = 1\frac{1}{10}$ **15** $5 - 2\frac{1}{7} = 2\frac{6}{7}$

How many $\frac{1}{2}$s are there in:

16 $2\frac{1}{2} = \frac{5}{2}$ **17** $4 = \frac{8}{2}$ **18** $3\frac{1}{2} = \frac{7}{2}$ **19** $5\frac{1}{2} = \frac{11}{2}$

20 $3 = \frac{6}{2}$ **21** $7\frac{1}{2} = \frac{15}{2}$

Write the following fractions as decimals.

22 $\frac{23}{100}$ _0.23_ **23** $\frac{7}{100}$ _0.07_ **24** $\frac{3}{10}$ _0.3_ **25** $\frac{9}{1000}$ _0.009_

26–30 Put these times in order, earliest first.

A 23.19   B 05.10   C noon   D 00.15   E 16.25

_D_   _B_   _C_   _E_   _A_

36

31 Add half 28 to twice 19. ....52....

32 Write in figures the number ten thousand and fourteen. ....10 014....

33 How many times can I take 12 from 324? ....27....

34 The distance all round a square is 3.4 m. How long is each side? ....85 cm....

35 A concert started at 19.30. $1\frac{1}{4}$ hours after this there was a ten-minute interval. When did the interval end? ....20.55....

36 If the second half of the concert is the same length as the first half, when does the concert end? ....22.10....

37 The sum of two numbers is 26, and their difference is 8. What is the larger number? ....17....

38 The smaller number is: ....9....

This chart shows how many marks 33 children got in a maths test.

39 How many children got 10 marks? ....1....

40 How many children got 9 marks? ....4....

41 How many children got 8 marks? ....3....

42 How many children got 7 marks? ....2....

43 How many children got 6 marks? ....5....

44 How many children got 5 marks? ....6....

45 How many children got 3 marks? ....2....

46 How many children got 2 marks? ....1....

47 How many children got 1 mark? ....2....

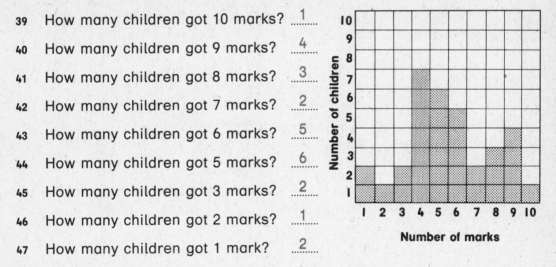

48 Fill in on the graph the number of children who got 4 marks.

49 What is the smallest number that can be divided by both 5 and 7 without a remainder?   35

50 It rained on 19 days in January. How many days were fine that month?   12

Paper 15

1–9 Make out a timetable for Class 3C. They have 4 lessons of 45 minutes each, with a break of 20 minutes in the middle of the morning.

	Begins		Ends
1st lesson	09.10	to	09.55
2nd lesson	09.55	to	10.40
Break	10.40	to	11.00
3rd lesson	11.00	to	11.45
4th lesson	11.45	to	12.30

10
$$\begin{array}{r} 9\,4 \\ 8\,\overline{)7\,5\,2} \end{array}$$

11 If Ian cycles at an average speed of 16 km/h, how long will he take to go 40 km? $2\frac{1}{2}$ hours

12 The product of two numbers is 56. One number is 7. What is the other?   8

Here is part of a centimetre ruler.

13 From P to R is   1.5 cm   **14** From P to T is   4 cm

15 From Q to U is   5 cm   **16** From R to U is   4.5 cm

17 From S to U is   3.5 cm   **18** From T to V is   2.5 cm

19	20	21	22

19   5 6
17) 9 5 2

20   9 1 9 1
    − 1 7 1 7
      7 4 7 4

21   1 4 5
    ×   3 0 8
      4 4 6 6 0

22 £
      7 8 . 0 5
       9 . 9 2
    + 5 . 0 7
      9 3 . 0 4

Turn to improper fractions:

23–26 $3\frac{1}{3}$ $\frac{10}{3}$ .... $4\frac{1}{2}$ $\frac{9}{2}$ .... $1\frac{3}{4}$ $\frac{7}{4}$ .... $1\frac{1}{6}$ $\frac{7}{6}$ ....

27–30 $2\frac{1}{5}$ $\frac{11}{5}$ .... $4\frac{2}{3}$ $\frac{14}{3}$ .... $5\frac{1}{2}$ $\frac{11}{2}$ .... $1\frac{5}{7}$ $\frac{12}{7}$ ....

31 What is the area of this garden? ....600 m²

32 What fraction of the garden is used for growing vegetables? $\frac{1}{3}$ ....

33 What fraction of the garden is the rose bed? $\frac{1}{6}$ ....

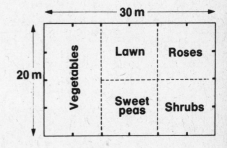

34 What fraction of the garden is the shrubbery? $\frac{1}{6}$ ....

35 What is the area of the vegetable garden? ....200 m²

36 What is the area of the shrubbery? ....100 m²

37 What is the perimeter of the garden? ....100 m

38–41 Using the digits 5 2 and 1, make:

the smallest number 125   the smallest even number 152

the largest number 521   the largest even number 512

42–43 Draw the line of symmetry in each figure.

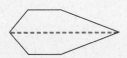

Write the following fractions as percentages.

44 $\frac{7}{10}$ = 70%　　**45** $\frac{10}{50}$ = 20%　　**46** $\frac{10}{20}$ = 50%　　**47** $\frac{30}{50}$ = 60%

48 $\frac{5}{20}$ = 25%　　**49** $\frac{20}{50}$ = 40%　　**50** $\frac{40}{50}$ = 80%

Paper 16

Write the next two numbers in each of the following lines:

1–2 77　　88　　99　　110　　121

3–4 $3\frac{1}{4}$　　$3\frac{1}{2}$　　$3\frac{3}{4}$　　4　　$4\frac{1}{4}$

5–6 987　　98.7　　9.87　　0.987　　0.0987

7
```
    m
  7.69
  4.82
+ 3.68
 16.19
```

8
```
    m
  4.12
- 2.76
  1.36
```

9
```
    m
  6.17
×    8
 49.36
```

10
```
      m
       3.49
8) 27.92
```

Here is a chart which shows the number of people who came to see our school play. ▨ represents 20 people.

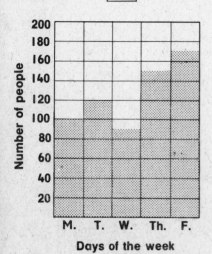

How many people came on:

11 Monday? 100

12 Tuesday? 120

13 Wednesday? 90

14 Thursday? 150

15 Friday? 170

16 The total audience was: 630

17 What is the area of a floor 2.5 m × 3.5 m?   8.75 m²

18 Take four hundred and thirty-nine from seven thousand
    five hundred.   7061

19 If the area of a room is 14 m² and the length is 4 m, what
    is the width?   3.5 m

Fill in the missing figures.

20 3 × 4 × □ = 24   21 8 ÷ 2 + ◯ = 10

    so □ = 2   so ◯ = 6

22 S + 7 + 3 = 12   23 4 × ◯ = 10 + 6

    so S = 2   so ⬡ = 4

A room has a perimeter of 18 m. It is twice as long as it is wide.

24–25 It is 6 m long and 3 m wide.

26–27 Divide these numbers by 10.

3.55	3.05	30.5	303.5	10
0.355	0.305	3.05	30.35	1

31 40 × 0.02 = 0.8   32 70 × 1.1 = 77

33 Add ¼ of 48 to twice 19.   50

34 How many times can I take 39 from 624?   16

35 How many minutes are there between 11.45 p.m. on
    Wednesday and 5.27 a.m. on Thursday?   342 min

	David	Stephen	Jerry	Simon
Maths	77	78	81	84
English	83	74	65	81
History	81	92	68	76
Science	69	68	49	77
	310	312	263	318

36–39 Add up each boy's marks.

40 Who had the most marks? .... Simon

41 Who had the lowest total marks? .... Jerry

42 Who had the highest marks in any one exam? .... Stephen

43 Who had the lowest marks in any one exam? .... Jerry

44–47 How many $\frac{1}{5}$ s are there in:

$1\frac{2}{5}$ $\frac{7}{5}$ $2\frac{1}{5}$ $\frac{11}{5}$ $1\frac{4}{5}$ $\frac{9}{5}$ $2\frac{2}{5}$ $\frac{12}{5}$

Class 1 have 200 beads in a box. $\frac{1}{5}$ are red, $\frac{3}{10}$ are blue and the rest are yellow.

48–50 There are ...60... blue beads, ...40... red beads and ...100... yellow beads.

Paper 17

1 How far is it from Red Cove to St. Aubrey? ...12 km...

2 How far is it from Ashville to Red Cove? ...11 km...

3 How far is it from Fingal Bay to St. Aubrey? ...13 km...

4 How far is it from Ashville to Fingal Bay? ...12 km...

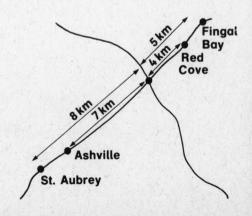

5-8 Underline the sums which have a remainder of 3.

$29 \div 5$ $21 \div 6$ $19 \div 4$ $28 \div 7$

$24 \div 7$ $77 \div 11$ $40 \div 12$ $27 \div 8$

9 If there are 20 lines on each page, on which page will the 110th line appear?   <u>6th page</u>

10 Robert has 52 marbles. He lost 16, and then won back twice as many as he had lost. How many has he now?   68

The 30 children in our class made this pictogram to show which are our favourite television programmes.

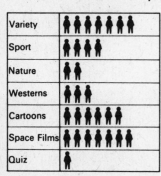

Variety	
Sport	
Nature	
Westerns	
Cartoons	
Space Films	
Quiz	

11 What fraction of the class prefers cartoons? $\frac{1}{5}$

12 How many children prefer space films?

13-14 What two programmes are equally popular?   <u>variety and space films</u>

15 What fraction of the class prefers nature films? $\frac{1}{15}$

16-17 Which two groups together number the same as those who like variety?

<u>cartoons and quiz or sport and westerns</u>

18 What fraction of the class prefers westerns? $\frac{1}{10}$

Double the following amounts.

19 £3.45   <u>£6.90</u>   **20** $3\frac{3}{4}$ kg   <u>$7\frac{1}{2}$ kg</u>

Halve the following amounts.

21 912   <u>456</u>   **22** £5.20   <u>£2.60</u>

23 I gave Tanya half of my chocolates, and had 37 left. How many had I at first?   74

24-27 Do the following sums in your head. Remember 99 is 1 less than 100 so you add on 100 and take away 1.

367 + 99	744 + 99	468 + 99	702 + 99
466	843	567	801

28 Take one hundred and seventy-seven from ten thousand. 9823

29 How many 24p stamps could I buy with £6.00? 25

30 In a class of 30 children one-tenth of them do not drink school milk. How many do drink it? 27

31 We bought 120 cakes for a party. 27 were left. How many were eaten? 93

32
$$7\overline{)319.2} = 45.6$$

33 $3\frac{1}{2} + 5\frac{1}{10}$ $8\frac{3}{5}$

34 Write the answer to Question 33 as a decimal. 8.6

35 What is the perimeter of a room 4.7 m long and 3.5 m wide? 16.4 m

36 How much less than 20 m is this? 3.6 m

37 Line A measures 1.5 cm   A _____

38 Line B measures 6.5 cm   B _____

39 Line C measures 3 cm   C _____

40 Line D measures 2.5 cm   D _____

If, on a map, 1 cm represents 20 km:

41 Line A would represent 30 km

42 Line B would represent 130 km

43 Line C would represent 60 km

44 Line D would represent 50 km

45–48 Write these times as you would see them on a digital clock.

Half past 7 in the morning _07.30_   Half past 8 at night _20.30_

Quarter to 11 in the morning _10.45_ Quarter past ten at night _22.15_

49 Share £6.65 equally among 19 boys.   _35p_

50 345 × 300   _103 500_

Paper 18

Write the following decimals as fractions in their lowest terms.

1 0.4 $\frac{2}{5}$ **2** 4.25 $4\frac{1}{4}$ **3** 9.9 $9\frac{9}{10}$

4 3.03 $3\frac{3}{100}$ **5** 0.01 $\frac{1}{100}$ **6** 10.07 $10\frac{7}{100}$

7 My aunt said that she would give my sister 4p for every kilogram she weighed. She weighed 19.5 kg. How much money did she receive?   _78p_

8 By how much is 3.5 greater than 0.6?   _2.9_

How many 1 cm squares are there in each shape?

9 7 squares

10 9 squares

11 9 squares

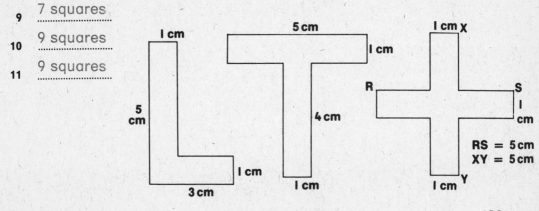

12 How many 14p stamps can I buy for £2.80?   _20_

Put a sign in each space to make each sum correct.

13–14 (6 ...×... 1) ...+... 2 = 8   **15–16** (4 ...+... 5) ...÷... 3 = 3

17–18 (7 ...−... 2) ...÷... 5 = 1   **19–20** (8 ...÷... 2) ...−... 4 = 0

Together Tim, Simon and Robin have 35 counters. Simon has half as many as Tim, and Tim has half as many as Robin.

21–23 Robin has ...20... counters, Simon has ...5... and Tim ...10...

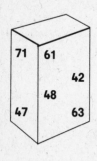

24 Which number can be divided exactly by 7 and 6? ...42...

25 Which number has a remainder of one when divided by 12? ...61...

26 Which number is 52 less than 100? ...48...

27 Which number has a remainder of 5 when divided by 11? ...71...

The 24 children in David's class made this pie-chart which shows their favourite pop groups.

28 How many like Jake and his Jokers best? ...9...

29 ...6... children like The Riders best.

30 Tommy Tucker is liked best by ...3... children.

31 The Bakers are liked best by ...6... children.

32 km	**33** m	**34** m	**35** km
9.420	4.68	6.54	9.500
− 7.860	9.25	× 9	+ 6.750
1.560	+ 3.79	58.86	16.250
	17.72		

36–39 Draw lines between sets of equal value:

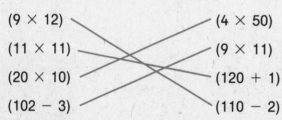

(9 × 12)   (4 × 50)

(11 × 11)   (9 × 11)

(20 × 10)   (120 + 1)

(102 − 3)   (110 − 2)

40 $\frac{7}{8} \times \frac{4}{14}$

$\frac{1}{4}$
....

41 $\frac{3}{4} \div \frac{1}{2}$

$1\frac{1}{2}$
....

42 $7 - 1\frac{1}{5}$

$5\frac{4}{5}$
....

43 $4\frac{11}{12} + 3\frac{1}{2}$

$8\frac{5}{12}$
....

Our class made this Venn diagram
to show who likes art best
and who likes P.E. best.

44 How many are
in our class? ....22...........

45–46 ....8.... prefer P.E. and
....3.... prefer art.

47 How many children
like both art
and P.E.? ....4.......

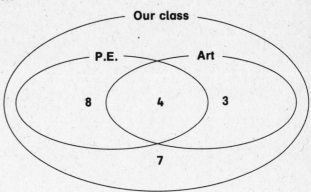

48–49 ...10... children don't like P.E. and ...15... don't like art.

50 If $\frac{3}{5}$ of my money is 21p, how much have I? ....35p...........

Paper 19

Complete these sums by filling in the missing figures.

1 $\frac{2}{3} = \frac{16}{24}$

2 $\frac{7}{10} = \frac{42}{60}$

3 $\frac{5}{9} = \frac{20}{36}$

4 $\frac{11}{12} = \frac{44}{48}$

5 $\frac{2}{7} = \frac{8}{28}$

6 $\frac{8}{9} = \frac{48}{54}$

7 Which two consecutive numbers add up to 17? .....8 and 9.....

8 Which two consecutive numbers add up to 25? .....12 and 13.....

9 How many seconds are there in 20 min 28 sec? .....1228.....

Here is a chart which shows the height of some
of my friends.

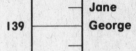
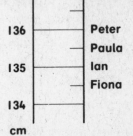

10 How much taller is Jane than Lucy? .....2 cm.....

11 How much taller is George
than Paula? .....3.5 cm.....

12 How much shorter is Ian than Lucy? .....2.5 cm.....

13 What is the difference in height
between Shaun and Fiona? .....2.5 cm.....

14 How much taller is Jane than Peter? .....3.5 cm.....

15 How much shorter is Paula
than Diana? .....2.5 cm.....

16 What is the difference between
the heights of Fiona and George? .....4.5 cm.....

17 How much taller is Jane than Ian? .....4.5 cm.....

18 0.25 m = ...25..... cm

19 0.5 kg = ...500..... g

20 1 km = ...1000..... m

21 346 cm = ...3.46..... m

22 12.2 cm = ...122..... mm

23 2.5 kg = ...2500..... g

24–25 Underline the larger of each pair.

4.7 metres/468 cm   22 minutes/ $\frac{2}{5}$ hour

26–27 The room is .....6 m..... long and
.....4 m..... wide.

28–29 The table is .....2 m..... long and
.....1 m..... wide.

30 The fireplace is .....1 m..... long.

31 How long is the window? .....2.5 m.....

32 How wide is the chair? .....50 cm.....

33 The area of the room is .....24 m².....

34 What is the perimeter of the room? .....20 m.....

Scale: 1 cm represents 1 m

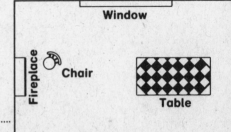

35 $3 \times \triangle + 2 = 26$

    so $\triangle = $ 8

36 $\square \times 2 - 4 = 18$

    so $\square = $ 11

37 $4 \times \bigcirc - 2 = 34$

    so $\bigcirc = $ 9

38 $2 \times \triangledown - 5 = 9$

    so $\triangledown = $ 7

Underline the correct answer in each line.

39	$\frac{1}{2} \times \frac{1}{2}$	$= \underline{\frac{1}{4}}$	$\frac{1}{8}$	$\frac{1}{2}$	1	2
40	0.1×0.1	$= 0.2$	0.02	0.1	<u>0.01</u>	0.11
41	$4 \div \frac{1}{4}$	$= 1$	4	<u>16</u>	2	$\frac{1}{4}$
42	$2 - 0.1$	$= 1.1$	<u>1.9</u>	2.1	0.9	1.2
43	$1 + 0.1 + 0.01$	$= 2.1$	1.2	2.01	1.101	<u>1.11</u>
44	$\frac{1}{2}$ of 66	$= 60$	66	30	<u>33</u>	25

45 If Tony can swim 15 m in 20 seconds, how far can he swim, at the same rate, in 2 minutes?

    90 m

46 If it takes Maria a quarter of an hour to walk to school, what time must she leave home to get to school by 8.50?

    8.35

47 How many books, each 1.2 cm thick, can be stood on a shelf 50.4 cm wide?

    42

48 What is the average (mean) of these numbers? 4 5 7 8

    6

49–50 In a school there are 200 pupils. $\frac{2}{5}$ of them are girls. There are 80 girls and 120 boys.

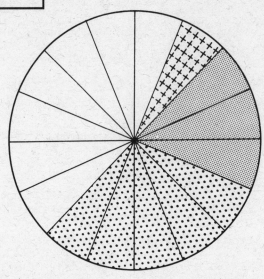

1–3 What fraction of the circle is:

dotted? $\frac{5}{16}$ shaded? $\frac{3}{16}$ plain? $\frac{7}{16}$

4 What fraction of the circle has crosses? $\frac{1}{16}$

5 What fraction of the circle is dotted and shaded? $\frac{1}{2}$

6 What fraction has crosses and dots? $\frac{3}{8}$

7 If I travel 150 km in $2\frac{1}{2}$ hours, what is my speed in km/h? 60 km/h

8 Which number is halfway between 17 and 29? 23

9–10 Look at the number 1212. What is the difference in the value of: the two 2 s? 198

the two 1 s? 990

11 Add half of 18 to twice 6. 21

12 Twice a number is 36. What is three times that number? 54

13 Laura spent four-fifths of her money. If she had 6p left, how much had she at first? 30p

14-19 Complete the following table:

Train departs at	Journey lasts	Train arrives at
09.30	40 minutes	10.10
09.42	30 minutes	10.12
10.45	40 minutes	11.25
11.15	50 minutes	12.05
11.40	40 minutes	12.20
12.35	30 minutes	13.05

20-22 What is the next number after 1738 which:

ends in a zero   1740

ends in 2 zeros   1800

ends in 3 zeros   2000

23 I think of a number, add 1, divide by 3, and the answer is 2. What is the number?

5

24 In my sister's box of counters half of them are green, one quarter of them are blue, and there are 11 white counters. How many are there altogether?

44

25 How long will it take a lorry to go 325 km at an average speed of 50 km/h?

$6\frac{1}{2}$ hours

26-27 The sum of the ages of Anita and Peter is 20 years. Anita is 2 years older than Peter. Anita is _11_ and Peter _9_

28 The difference between two numbers is 14. The smaller number is 17. What is the other?

31

29 In a village of 550 people $\frac{2}{5}$ of the population are adults. How many children are there?

330

30 In a test, 24 out of 30 passed. What fraction failed?

$\frac{1}{5}$

31 How many times can I take 5 from 125?

25

51

The correct time is 5 past 10. How fast or slow are these clocks? Write the number of minutes, and cross out the words not needed.

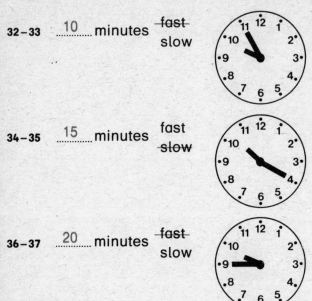

32–33   ...10... minutes   ~~fast~~
                            slow

34–35   ...15... minutes   fast
                            ~~slow~~

36–37   ...20... minutes   ~~fast~~
                            slow

38–47   Make out a timetable below for Brambleash Primary School. They have five 30 minute lessons each morning, with a break of 20 minutes after the second lesson.

	Begins		Ends
1st lesson	09.10	to	09.40
2nd lesson	09.40	to	10.10
Break	10.10	to	10.30
3rd lesson	10.30	to	11.00
4th lesson	11.00	to	11.30
5th lesson	11.30	to	12.00

48   How long do they spend at work each morning? $2\frac{1}{2}$ hours

49   I bought 7 toys at 65p each. How much did this cost?   £4.55

50   How much change would I have out of £5.00?   45p

Paper 21

Insert a sign in each space to make each sum correct.

1-2 3 $\times$ 4 = 13 $-$ 1

3-4 8 $+$ 1 = 3 $\times$ 3

5-6 10 $\div$ 2 = 4 $+$ 1

7
```
days  hr
  4    7
×      5
──────────
2 1 1 1
```

8
```
      4 7 9
1 1 ) 5 2 6 9
```

9
```
  7 0 0 1
- 1 0 0 7
─────────
  5 9 9 4
```

10
```
         4 8
2 8 ) 1 3 4 4
```

11 Write in figures the number: one hundred and one thousand.   101 000

12 What number, when divided by 11, has an answer of 17 remainder 2?   189

The sum of two numbers is 45, and their difference is 7.

13-14 The smaller number is   19   and the larger   26

15
```
    4 7 6
  × 5 0 0
─────────
2 3 8 0 0 0
```

16
```
    0 . 6 2 4
7 ) 4 . 3 6 8
```

17
```
      hr  min
       6    6
1 1 ) 6 7    6
```

18
```
min  sec
  8   22
- 6   30
─────────
  1   52
```

19 I think of a number, divide it by 4 and then add 2. The answer is 6. What is the number?   16

Complete the following table.

20-23

	Length	Width	Perimeter
Rectangle 1	7 m	5 m	24 m
Rectangle 2	6 m	4 m	20 m
Rectangle 3	8 m	3 m	22 m
Rectangle 4	9 m	4 m	26 m

The children in our class made this chart which shows our favourite lessons.

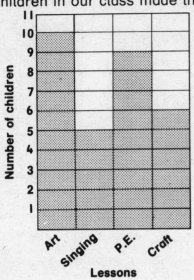

24	How many children are there in the class?	30
25	Which is the most popular lesson?	Art
26	Which is the next most popular?	P.E.
27	How many more children prefer craft than prefer singing?	1
28	There are 4 more boys than girls in our class. How many boys are there?	17
29	How many girls are there in the class?	13
30	My watch gains 5 minutes a day. If I put it right at 6 p.m. on Sunday, what time will it show at 6 p.m. on Tuesday?	6.10
31	At 6 p.m. on Thursday?	6.20
32	At 6 p.m. a week on Sunday?	6.35
33	Take 22 cm from 13 m.	12.78 m
34	Add together 9 kg, 276 g and 14.5 kg	23.776 kg

Put a decimal point in these numbers so that the 1 will have the value of 1 unit.

4 3 1·2   1·2 4 3   2 1·3 4   3 4 2 1·   1·0 4 3

Our class made this chart, which shows our favourite pets.

□ represents two children.

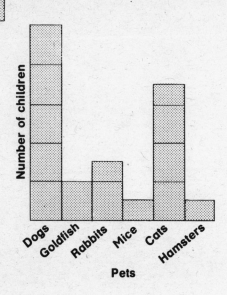

40 How many are there in our class? ....24........

41 What fraction of the class prefers rabbits? ....$\frac{1}{8}$........

42 How many children prefer dogs? ....10........

43 How many children like hamsters best? ....1........

44 How many more children like dogs than like cats? ....3........

45 What fraction of the class prefers goldfish? ....$\frac{1}{12}$........

46-47 Which two pets are equally popular? ....hamsters and mice........

48-49 Which two groups put together equal those who prefer dogs? ....cats and rabbits........

50 How many more children prefer cats than prefer rabbits? ....4........

Write the next three items in each of the following lines:

1–3	12	14	15	17	18	20	21	23
4–6	$\frac{7}{8}$	$\frac{3}{4}$	$\frac{5}{8}$	$\frac{1}{2}$	$\frac{3}{8}$	$\frac{1}{4}$		
7–9	60	59	49	48	38	37	27	26

10–12 May 31   June 30   July 31   August 31   September 30

13–15	100	96	92	88	84	80	76

The children in Brambledown Village School made this chart which shows their ages.

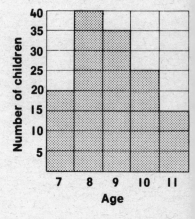

16 How many children are there in the school? 135

17 Which age group is the largest? 8-year-olds

18 Which age group is the smallest? 11-year-olds

19 How many more children are there who are 8 than are 9? 5

20 If half the children who are 7 are boys, how many girls are 7 years old? 10

21 If there are 26 8-year-old boys, how many girls are 8? 14

22 If there are 20 9-year-old girls, how many boys are 9? 15

23 If the boys outnumber the girls by one in the 10-year-old class, how many boys are there? 13

24 How many girls are 10 years old? 12

25 There are twice as many girls as boys in the 11-year-old class. How many girls are there? 10

26 How many boys are 11? 5

27 Jane spent $\frac{1}{4}$ of her pocket money on a book, saved $\frac{1}{2}$ of it, and spent the rest on some chocolate. If she spent 75p on the chocolate, how much had she at first? ....£3.00....

28–32 Arrange these decimals in order, putting the largest first.

2.02	2.2	2.222	2.12	2.22
2.222	2.22	2.2	2.12	2.02

33 km
9 . 2 5 4
− 7 . 3 8 6
─────
1 . 8 6 8

34 kg
1 4 . 3 7 9
+ 1 1 . 7 0 4
─────
2 6 . 0 8 3

35 min sec
6   1 2
×   6
─────
3 7   1 2

36 £
3 . 2 6
2 1 ⟌ 6 8 . 4 6

37 What sum is 14p more than 37p? ....51p....

38–42 Write these times as you would see them on a digital clock.

A quarter to 12 in the morning ....11.45....

20 to 11 at night ....22.40....

10 to 1 in the middle of the day ....12.50....

25 to 12 at night ....23.35....

5 minutes to 7 in the morning ....06.55....

43–47 Can you match these fractional parts and whole numbers? Join them with lines.

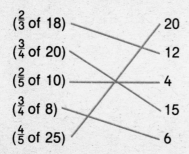

($\frac{2}{3}$ of 18)   20

($\frac{3}{4}$ of 20)   12

($\frac{2}{5}$ of 10)   4

($\frac{3}{4}$ of 8)   15

($\frac{4}{5}$ of 25)   6

48 The train, due at 11.53, was 11 minutes late. When did it arrive?   12.04

49 Add $\frac{1}{5}$ of 10 to twice 24.   50

50 Kirstie had 50 marbles. She lost 12 and then won back 3 times as many as she had lost. How many has she now?   74

Paper 23

1–3 Complete the fractions.

$\frac{3}{6} = \frac{4}{8}$ $\frac{5}{5} = \frac{7}{7}$ $\frac{2}{10} = \frac{4}{20}$

4 Take 11 from 11 thousand   10 989

5 If Sparvux costs £2.80 for 5 litres, how much should I pay for $\frac{1}{4}$ litre?   14p

6 When I take 4 times a certain number from 50 the remainder is 18. What is the number?   8

7 The sum of two numbers is 43. The smaller number is 17. What is the other number?   26

8–10 abbrabbr   What would the 11th letter be?   b

bytobyb   What would the 10th letter be?   o

xxyzzxx   What would the 16th letter be?   x

11 What is the perimeter of a room which is 5·4 metres long and 3·97 metres wide? Answer in metres.   18·74 m

12 Take 347 centimetres from 5.2 metres. Answer in centimetres.   173 cm

13–18 Put a decimal point in each of the following numbers so that the numeral 5 will have the value of 5 units.

7 5·6   5·7 6   7 6 5·   6 7 5·   5·6 7   6 5·7

Two trains leave Middleton at 10.30 a.m., one travelling non-stop due north at 120 km/h, and the other travelling non-stop due south at 100 km/h.

19 By midday the first train will have gone 180 km.

20 By midday the second train will have gone 150 km.

21 The distance between the two trains will be 330 km at midday.

22 By 1 p.m. the first train will have travelled 300 km.

23 By 1 p.m. the second train will have travelled 250 km.

24 So now they will be 550 km apart.

25 Divide 86.6 by 200   0·433

26 0.2 × 0.2 × 0.2 =   0·008

27 Find the product of 3.46 and 0.5   1·73

28 How many times can I subtract 0.06 from 18.6?   310

29
```
    8 7
    6 5
    4 6
    3 8
  + 2 5
   ────
    2 6 1
```

30 The train, which was 13 minutes late, arrived at 18.02. When should it have arrived?   17.49

31 The perimeter of the church hall is 72 m. It is twice as long as it is wide. What is its length?

24 m

32 What is its width?

12 m

33 There are 8 brown dogs and 4 white ones. What fraction of the dogs are white?

$\frac{1}{3}$

34 There are 28 children in a class. $\frac{3}{7}$ of them are boys. How many boys are there?

12

35 How many girls are there?

16

36–40 Round off to the nearest 100.

234	478	351	770	191
200	500	400	800	200

41 Three numbers are multiplied together to make 360. Two of the numbers are 12 and 10. What is the third number?

3

42 What is the sum of the odd numbers between 8 and 16?

48

43 From the product of 9 and 6 take their difference.

51

44 How many minutes are there between 10.49 and 11.32?

43 min

45 In our class 12 girls have short hair, and 4 girls have long hair. What fraction of the girls have long hair?

$\frac{1}{4}$

Belford-on-Sea has a population of 2015. The men and women together number 1007, and the women and children together number 1537.

46 How many children are there?

1008

47 How many women are there?

529

48 How many men are there?

478

60

49 The ages of Alison and Gavin add up to 26 years. Alison is 2 years older than Gavin so Alison is .... 14 years

50 Gavin is .... 12 years

Paper 24

Fill in the missing times:

	Train departs	Journey lasts	Train arrives
1	13.30	50 minutes	14.20
2	13.50	55 minutes	14.45
3	14.55	30 minutes	15.25
4	15.05	35 minutes	15.40
5	15.25	45 minutes	16.10
6	16.30	45 minutes	17.15

7 How many days are there in 45 weeks? 315

8–10 Write down all the numbers between 20 and 50 that are exactly divisible by 9. .... 27, 36, 45

Fill in the two missing numbers in each line.

11–12	12	16	20	24	28		32
13–14	35	28	21	14	7		
15–16	24	36	48	60	72		84
17–18	9	$7\frac{1}{2}$	6	$4\frac{1}{2}$	3		$1\frac{1}{2}$
19–20	777	77·7	7·77	0·777	0·0777		
21–22	81	27	9	3	1		

61

23 How many minutes are there between 11.36 and 15.43?   247 min

24
```
  7 5 6
×     9
  6 8 0 4
```

25 m
```
    4·8 8
×       7
  3 4·1 6
```

26
```
        9 1
11 ) 1 0 0 1
```

	Train A	Train B	Train C
Middlesbrough	11.35	13.20	14.35
Stockton	11.46	13.44	14.46
West Hartlepool	12.07	14.07	15.07
Sunderland	12.36	14.36	15.36
Newcastle	13.04	15.05	16.03

27 How long does train A take to travel from Middlesbrough to Newcastle?   1 hr 29 min

28 How long does train B take to do the same journey?   1 hr 45 min

29 Which is the fastest of the three trains?   Train C

30 How long does train C take to travel from Sunderland to Newcastle?   27 min

31 How long does train A take to travel from Middlesbrough to West Hartlepool?   32 min

32 How long does train B take to travel from Stockton to West Hartlepool?   23 min

33 How long does train A take to travel from Stockton to Sunderland?   50 min

34 How many minutes more does train B take to travel from Stockton to Sunderland than train C?   2 min

A mixed number is a whole number together with a fraction, such as $1\frac{1}{2}$.

Write these improper fractions as mixed numbers:

35   $\frac{7}{5} = 1\frac{2}{5}$ 36 $\frac{9}{8} = 1\frac{1}{8}$ 37 $\frac{3}{2} = 1\frac{1}{2}$

38   $\frac{5}{3} = 1\frac{2}{3}$ 39 $\frac{8}{5} = 1\frac{3}{5}$ 40 $\frac{9}{2} = 4\frac{1}{2}$

41   Write $\frac{22}{100}$ as a decimal.   0·22

42   How many quarters are there in $4\frac{1}{2}$?   18

43–46   Fill in the missing numbers:

$6 \times 7 \times \underline{10} = 420$ $2 \times 3 \times \underline{50} = 300$

$3 \times 4 \times \underline{30} = 360$ $5 \times 4 \times \underline{40} = 800$

47–50

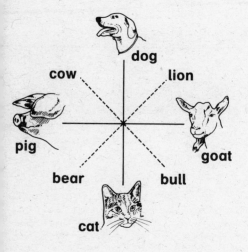

Start at dog and move 3 places clockwise.   bull

Start at cat and move 2 places anticlockwise.   goat

Start at pig and move 5 places anticlockwise.   lion

Start at bear and move 7 places clockwise.   cat

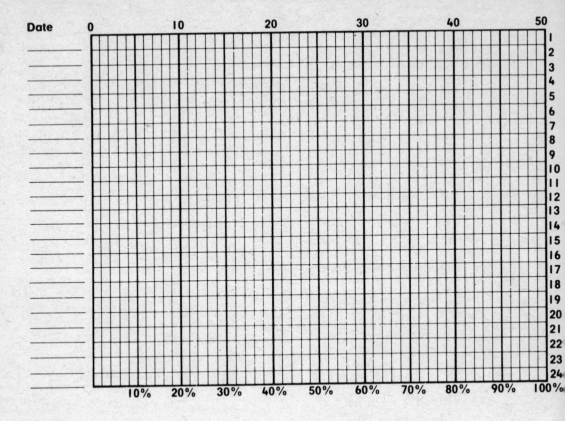

Thomas Nelson and Sons Ltd
Nelson House Mayfield Road
Walton-on-Thames Surrey
KT12 5PL UK

Thomas Nelson Australia
102 Dodds Street
South Melbourne
Victoria 3205 Australia

Nelson Canada
1120 Birchmount Road
Scarborough Ontario
M1K 5G4 Canada

I(T)P Thomas Nelson is an International
Thomson Publishing Company

I(T)P is used under licence

© **J M Bond** 1973, 1983, 1986, 1987, 1994
First published by Thomas Nelson and Sons Ltd 1973
Second edition 1983
Revised edition 1986
This fully revised edition 1994
Reprinted 1995

Pupil's book   ISBN 0-17-424505-X
   NPN 9 8 7 6
Answer book ISBN 0-17-424506-8
   NPN 9 8 7 6

By the same author
First, Second, Fourth and Further Fourth Year
Assessment Papers in Mathematics

First, Second, Third, Fourth and Further Fourth Year
Assessment Papers in English

First, Second, Third, Fourth and Further Fourth Year
Assessment Papers in Reasoning

Printed in Croatia.